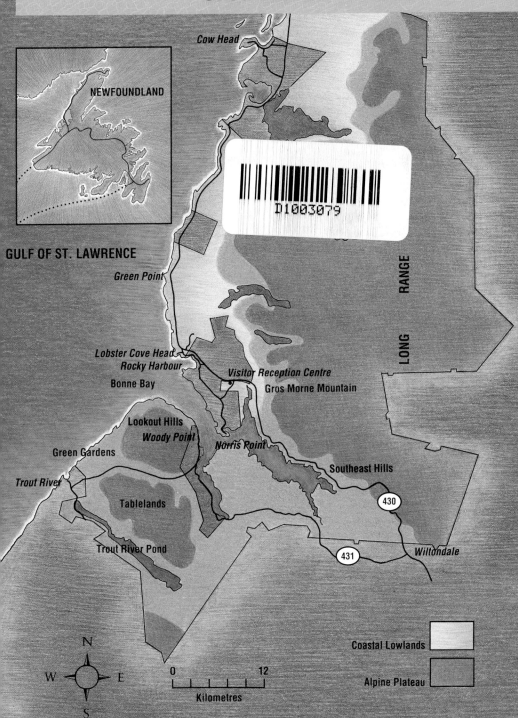

Gros Morne National Park

NEWFOUNDLAND

GULF OF ST. LAWRENCE

Cow Head

Green Point

LONG RANGE

Lobster Cove Head
Rocky Harbour
Bonne Bay

Visitor Reception Centre
Gros Morne Mountain

Lookout Hills
Woody Point

Norris Point

Green Gardens

Southeast Hills

Trout River

430

Tablelands

Trout River Pond

431   Wiltondale

N
W   E
S

0   12
Kilometres

Coastal Lowlands

Alpine Plateau

Gros Morne Co-operating Association
Rocky Harbour • Newfoundland • A0K 4N0 • Canada

Acknowledgements
A book such as this requires the help of many people. We would like to thank Karole Pittman for project co-ordination, Cathy Tutton and David MacDonald for design, and Kevin Sollows for illustrations. We are also very grateful to geologists Dr. H. Williams and Dr. R.K. Stevens, formerly at Memorial University of Newfoundland, and Dr. A.R. Berger, formerly with the Geological Survey of Canada, for their careful review and critique of the text.

Written by Michael Burzynski and Anne Marceau.
Imagesetting by Braemar Publishing, Halifax.
Photoengraving by Owen Innes Lithographic.
Printed in Canada by McCurdy Printing Limited.

Photographs
Cover: The Tablelands, ©James Steeves/Atlantic Stock Images.
Pages 4-5: The entrance to Bonne Bay, by James Steeves.
Pages 10-11: Sea stacks at Green Gardens, ©James Steeves/Atlantic Stock Images.
Pages 18-19: View from the Lookout Hills, by James Steeves.
Pages 30-31: Atop Gros Morne, ©James Steeves/Atlantic Stock Images.

Second edition, published by the Gros Morne Co-operating Association.
Copyright ©1990 Minister of Supply and Services Canada.
Reproduced with the permission of the Minister of Supply and Services Canada, 1995.

The publisher gratefully acknowledges Parks Canada, Department of Canadian Heritage as the source of this material.

Canadian Cataloguing in Publication Data
Burzynski, Michael, 1954-

Rocks Adrift

Includes bibliographical references and index.
ISBN 0-9699509-0-X

1. Geology --Newfoundland -- Gros Morne National Park -- Guidebooks.   2. Gros Morne National Park (Nfld.) -- Guidebooks.   I. Marceau, Anne, 1956- II Gros Morne Co-operating Association. III Title.

QE199.B87 1995   557.18   C95-950118-5

Contents

A World Heritage Site

GROS MORNE NATIONAL PARK WAS
established as a permanently protected site
by the Government of Canada in 1973.
The park is a wild landscape of fjords,
seacoast, forest, and mountains, but it was
geology that led to Gros Morne's
designation as a site of world-wide natural
significance. UNESCO, the United Nations
Educational, Scientific, and Cultural
Organization, declared the park a World
Heritage Site in 1987.

DURING THE PAST 1,250 MILLION YEARS, THE ROCKS THAT underlie Gros Morne National Park took shape during a series of events that changed the face of the Earth. Our planet is so large in comparison to us that we think of its stony surface as rigid, timeless, and unyielding, but at Gros Morne the rocks tell a tale of dramatic and continuous change.

Key stages of the Earth's history are recorded in park rocks. In Gros Morne, geological features that usually lie tens of kilometres underground are exposed at the surface. Some features are so unusual that they are nationally and even internationally famous. Park cliffs and outcrops provide evidence for one of the most important concepts in modern science: *plate tectonics*.

Plate tectonic theory suggests that continents are not fixed in place, but wander across the surface of the Earth at about the same speed that a fingernail grows. Plate tectonics offers a unified explanation for earthquake belts, mountain formation, volcanic activity, the shape of continents, the presence of mid-ocean ridges, and the world-wide distribution of minerals.

This theory explains the movements and collisions of continent-sized plates of rock during vast periods of time. It describes the very birth and death of oceans and the amalgamation and separation of continents. The unusual geology of the Gros Morne region has made the park an outdoor laboratory for the study of this earth-shaking concept.

Biologists too find the concept of plate tectonics useful — it makes sense of the distribution of fossil and contemporary plants and animals. Plate movements have influenced plant and animal evolution, and may indirectly have caused some of the great extinctions of the past.

The Shape of the Land

Gros Morne National Park protects 1,805 square kilometres of the Great Northern Peninsula of western Newfoundland. It comprises two main land types: an alpine plateau, and a coastal lowland sandwiched between the slopes of the Long Range Mountains and the waters of the Gulf of St. Lawrence. Around the inner arms of Bonne Bay are hills of intermediate height.

© James Steeves

Two Plains

Ancient rocks of the Long Range Mountains form an alpine plateau more than half a kilometre above the coastal lowland.

Bottom right: Ice-scoured terrain of the alpine plateau.

Rising as a series of cliffs and steep hillsides, the Long Range Mountains tower as much as 800 m above sea level. Atop the Long Range is the vast alpine plateau: a rolling landscape of ponds and weathered rock, covered in a patchwork of stunted forest, sub-arctic heaths, and grassy fens.

Gros Morne's coastal lowland lies just above sea level. Forced beneath the sea by the weight of glaciers during the Ice Age, the land rebounded as the ice melted. Deep fjords carved into the Long Range Mountains were blocked from the sea as the coastal lowland rose above salt water. These long narrow lakes are known locally as *ponds*.

Peat bogs blanket the poorly drained coastal lowland, and forests of fir and spruce cling to rocky ridges and the base of mountain slopes. Steep headlands, sea caves, broad cobble beaches, sea stacks, and dunes of fine sand line the shore.

James Steeves

The Earth in Action

Anatomy of the Earth

A SLICE THROUGH THE EARTH WOULD REVEAL THREE MAIN layers: the core, the mantle, and the crust.

At the very centre of the Earth is its *core*, a sphere about 7,000 km in diameter. The thousands of kilometres of overlying rock exert tremendous pressure on the molten nickel and iron of the core, and its temperature is estimated to be more than 2,700 °C.

Surrounding the core is the *mantle*, a 3,000-km-thick layer of dense magnesium- and iron-rich rock called *peridotite*. Heat generated by radioactive decay and gravitational friction keeps part of the mantle soft. Convection currents stir this semi-fluid layer. Molten rock (*magma*) rises towards the surface of the Earth, cools, and sinks to be reheated far below. On a smaller scale, convection currents can be seen in a cup of hot coffee as

Rock Types

Michael Wood

Igneous rocks are crystalline or glassy rocks formed by the cooling of molten material below or at the surface of the Earth. These are the primary source of the Earth's crust.

Examples: granite (shown above, about twice actual size), basalt, gabbro, peridotite.

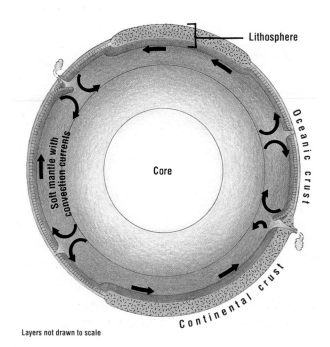

Layers not drawn to scale

Sedimentary rocks are cemented accumulations of fragments of other rocks, including deposits of plant and animal remains.

Examples: sandstone, shale, conglomerate, breccia, and limestone (shown above, reduced to about 1/10th actual size).

Metamorphic rocks are igneous or sedimentary rocks altered by heat and/or pressure and/or chemically active fluids within the Earth's crust.

Examples: gneiss (shown above, actual size), schist, and serpentinite.

milk warms and wells upwards in billows, then cools and sinks again.

Early in the Earth's history, light minerals floated to the surface of the molten planet. They cooled and hardened into a thin rigid *crust.* Mountain chains and ocean basins are mere wrinkles and sags in this cool hard shell. The crust ranges from 5 to 70 km thick. If Earth was the size of an apple, its crust would be thinner than an apple's skin.

Together the crust and solid upper mantle comprise the Earth's *lithosphere,* afloat on the soft mantle below.

The Earth has two main types of crust:

Continental crust ranges from 10 to 70 km in thickness, and is a mix of granites and other igneous rocks with a high content of relatively light-weight, light-coloured, silica-rich minerals. It also includes sedimentary and metamorphic rocks derived from igneous rocks and organic sources.

Oceanic crust is only 5 to 8 km thick, and is formed from dark, heavy, igneous rocks such as basalt and gabbro. *Basalts* are fine-grained rocks produced when molten rock cools and hardens quickly at the Earth's surface. *Gabbros* are coarse-grained rocks chemically equivalent to basalt, but cooled slowly underground. Basalts and gabbros have a high iron and magnesium content and are denser than continental crust rock. However, none of the crustal rocks are as dense as mantle rock.

Continental crust is relatively light, and floats higher on the soft mantle than does oceanic crust. Between continents, the thin heavy oceanic crust sags — producing basins in which the Earth's oceans have accumulated.

Plates Adrift

CURRENTS IN THE SOFT MANTLE ARE POWERFUL ENOUGH TO crack Earth's crust and upper mantle rock into enormous rafts, or plates. Along the cracks new crust forms as mantle material wells up to heal the gaps. The great slabs of lithosphere bump into, grind against, and override one another as they float on the slowly-churning molten mantle.

Since most earthquakes occur where plates meet each other, mapping tremors has helped scientists to locate plate edges. At present there are about twenty plates of different sizes and shapes moving about the surface of the planet. Some underlie ocean, while others underlie ocean and continent.

Earth's Fractured Surface
As plates spread apart, new ocean floor is created. When they converge, edges buckle into mountain chains, and earthquakes and volcanos result.

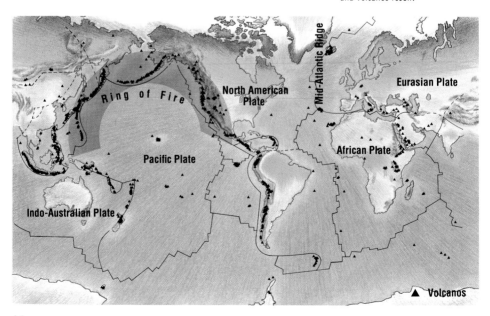

Ring of Fire

North American Plate

Mid-Atlantic Ridge

Eurasian Plate

Pacific Plate

African Plate

Indo-Australian Plate

▲ Volcanos

Plate tectonics is a complex process involving changes in the size, shape, location, and number of plates. The Earth is about 4,500 million years old. Plate movements have occurred for at least 3,500 million years, probably since the crust first hardened. Driven by convection currents in the mantle, plates can move at the dizzying speed of ten centimetres a year.

When a continental plate collides with an oceanic plate, the dense oceanic plate is usually forced underneath the lighter continental plate, a process called *subduction*. Today, Pacific Ocean floor is being swallowed beneath surrounding continents, causing volcanos and earthquakes and creating volcanic *island-arcs* such as the Japanese Archipelago and the Aleutians. On a global scale, the mighty Pacific is a shrinking ocean.

The Ring of Fire that surrounds the Pacific includes such famous volcanos as Mount Fuji, Mount Mayon, and Krakatoa. The recent eruption of Mount St. Helens in the U.S. state of Washington was a result of the edge of the Pacific Plate melting deep beneath the west coast of North America, and the new magma melting its way to the surface.

When light continental plates collide, usually neither will sink. Instead, they buckle at their leading edges like cars in an accident. Masses of continental crust are thrust up, twisted, and folded, while absorbing the immense force of the collision. In this way the great mountain ranges of the world are created.

Plate movements are slow. The lofty Himalayas are still being pushed upward by a collision that began when the Indo-Australian Plate nudged against the Eurasian Plate some 50 million years ago. The Indo-Australian Plate is still advancing, and Mount Everest grows ever higher.

Continents endure by floating on the surface of the mantle, but ocean crust is consumed by subduction. Because ocean floor rock is continuously created at mid-ocean spreading centres, it is far younger than most continental rock. The oldest continental rocks in Newfoundland and Labrador are 3,800 million years old, but the oldest rocks in the ocean are only 150 million years old.

When Plates Move Apart

When plates move apart they create an immense crack, or rift. As the plates separate, molten rock wells into the expanding rift, then cools and hardens into new oceanic crust.

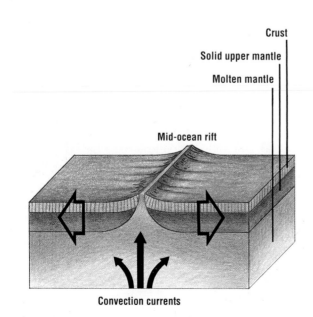

Crust

Solid upper mantle

Molten mantle

Mid-ocean rift

Convection currents

When Plates Collide

When plate edges override one another, one of the plates is forced down into the hot mantle and melts. This process is called subduction. Molten crust material is lighter than mantle, and it rises — melting its way through the overlying solid rock and erupting as volcanic lava.

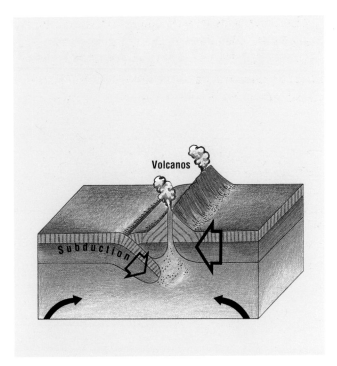

Volcanos

Subduction

When Plates Sideswipe

When plates slide past each other, the jolts and shudders are felt as tremors. This occurs at plate boundaries called transform faults. California's quake-ridden San Andreas Fault is a transform fault. There, the Pacific Plate moves northward against the North American Plate. Mid-ocean spreading centres are broken into zigzags by numerous transform faults.

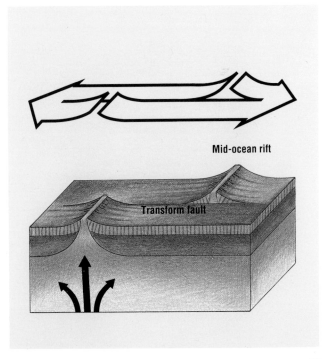

Mid-ocean rift

Transform fault

Gros Morne Assembled

An Ocean Created

THE LONG RANGE MOUNTAINS ARE THE BACKBONE OF GROS
Morne National Park, and contain the oldest rock on the
Island of Newfoundland. Dating back 1,250 million years,
they represent the eastern edge of the massive Canadian
Shield. Over time, heat and pressure transformed the
granitic Shield rock into gneiss and schist — banded
metamorphic rocks. About 1,000 million years ago,
magma melted its way into this rock and solidified as
granite.

The Canadian Shield and all of the Earth's other
landmasses were part of a super-continent. About 650
million years ago, the super-continent began to break
apart. Currents in the soft mantle wrenched at the
overlying rock until it stretched and subsided. Finally a
crack rent the lithosphere to the east of what is now Gros
Morne. Molten rock oozed up into cracks around the rift
and hardened as sheet-like vertical dykes.

By 600 million years ago the two landmasses had
moved apart, creating a rift valley similar to the Red Sea

Rifting of a Super-continent

Gros Morne's oldest rocks record
the break-up of a super-continent.
A rift split the landmass, then
widened and created a rift valley.
Erosion filled the valley bottom with
sediment.

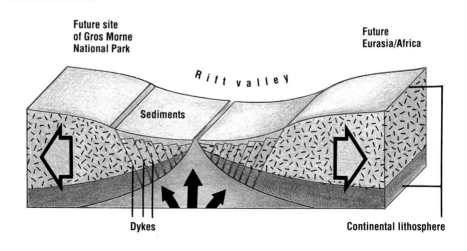

Future site
of Gros Morne
National Park

Future
Eurasia/Africa

Rift valley

Sediments

Dykes

Continental lithosphere

and African rift valleys of today. Magma surged up through the torn bedrock, cooling and hardening at the surface. Water collected as the basin expanded, eventually producing the Iapetus Ocean. Beneath the waters of the Iapetus, along the margin of a newly formed continent, were the rocks that would become Gros Morne National Park.

For a hundred million years sediments eroded from the land. Washed offshore, they accumulated as a *continental shelf* that extended 100 km out beneath the sea. The climate along this ancient coast was tropical, and the seas teemed with life. Calcium-rich mounds of coral-like algae and the skeletons and shells of trilobites, primitive corals, brachiopods, molluscs, and other organisms blanketed the sea floor.

Birth of an Ocean

The new plates pulled apart, and an ocean called the Iapetus developed between them. A wide continental shelf grew as sediments settled to the sea floor, and the shelly remains of sea creatures piled up as a thick carbonate (limestone) bank.

In time, these shelly remains built up a kilometre-thick *carbonate bank* that extended to the edge of the continental shelf. As the growing bank hardened into limestone, its steep outer face collapsed in periodic undersea avalanches. Fragments of limestone tumbled down the continental slope to the deep sea floor. Large pieces sank quickly, while clouds of finer particles settled

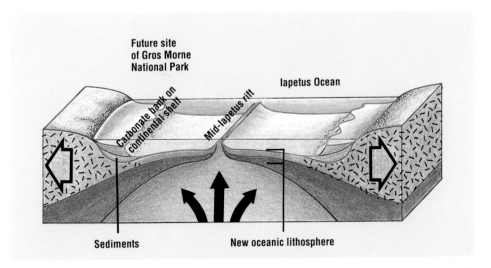

Future site of Gros Morne National Park

Carbonate bank on continental shelf

Mid-Iapetus rift

Iapetus Ocean

Sediments

New oceanic lithosphere

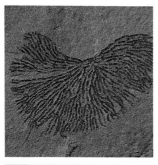

1

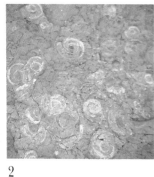

2

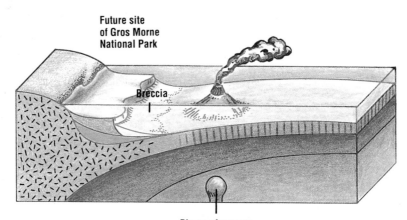

3

Animals from the Past

Fossils of marine life from warm shallow seas reflect the climate that prevailed along the continental shelf 500 million years ago.

1. graptolite colony
2. gastropods
3. trilobite

Avalanches and Volcanos

Occasionally, the edge of the carbonate bank crumbled, sending avalanches of limestone to the sea floor far below. These undersea landslides later hardened into a chunky natural concrete called breccia. Far offshore, volcanos were active over a hot-spot in the Earth's mantle.

Future site
of Gros Morne
National Park

Breccia

Plume of magma

more slowly, filling spaces between the chunks. Dust-like layers of mud and the remains of deep sea organisms, including the tiny, branched, colonial animals called *graptolites,* drifted down onto the limestone jumble in the long periods between avalanches. Eventually the coarse limestone mixture cemented together as *breccia.* The mud became a dark finely-layered *shale.*

By 500 million years ago, volcanic islands called *seamounts* had risen from the ocean floor several hundred kilometres offshore. Far from any subduction zone, these volcanos grew out of the ocean floor over a hot-spot in the Earth's mantle — like today's Hawaiian volcanos. Their lava came from a depth of more than 100 km.

The Iapetus had become a wide ocean, but in the time scale of plate tectonics, even an ocean is not permanent.

The Cow Head Breccia

Limestones and shales at Cow Head record a long series of undersea avalanches and contain fossils of both shallow- and deep-sea organisms.

James Steeves

An Ocean Destroyed

BETWEEN 500 AND 450 MILLION YEARS AGO, PLATE movement shifted into reverse. The continents on either side of the Iapetus Ocean began to move together again.

During compression, plates usually rupture near the edge of a continent. This time however, the ocean floor between the continents cracked far out to sea. Although the next stages are not completely understood, geologists agree that as the Iapetus narrowed, the ocean floor of the eastern plate (Eurasia and Africa) slid over the undersea edge of the plate that now forms North America.

This forced the edge of the North American Plate down towards the mantle. Melted crust from the descending slab bubbled to the surface, producing an arc of volcanic islands several hundred kilometres from shore (remnants can be seen today around Notre Dame Bay in central Newfoundland).

The converging continents thrust slices of deep-sea sediments, limestone breccia, and the carbonate bank into stacks. Even the stacks were sliced and re-stacked. Finally, slices of oceanic crust and mantle were pushed westward on top of the sedimentary rocks. The layers on which the rocks slid were crushed into a *mélange* of blocks embedded in a shaly ground-mass. Movement ceased when the roots of the island-arc volcanos jammed against the edge of the continent.

Closing of an Ocean

Plate movement reversed and the Iapetus began to close. The eastern plate overrode the western plate, and an arc of volcanic islands rose above the subduction zone.

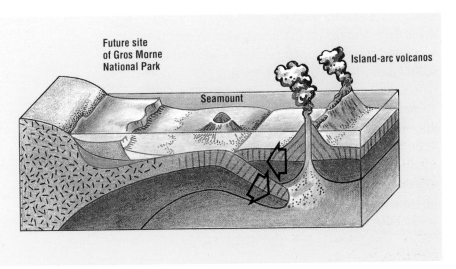

Future site of Gros Morne National Park

Seamount

Island-arc volcanos

Ocean Floor Stranded on Land

As the continents converged, the rocks in between were stacked one upon another, like ice floes on a shore.

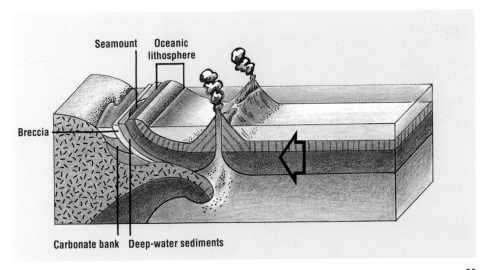

Seamount

Oceanic lithosphere

Breccia

Carbonate bank

Deep-water sediments

The two continents first made contact in the north, then squeezed together farther and farther south — like a closing zipper. This collision between North America and Eurasia/Africa began about 450 million years ago and took 100 million years to complete. It crumpled the edge of the western continent into a 4,000 km mountain chain as rugged as the Rockies. This Caledonide Mountain System ran north through what are now Great Britain and Scandinavia, and extended south to form North America's Appalachian Mountains.

After more than 100 million years as an ocean, the Iapetus was gone. Reunification of the landmasses produced a new super-continent — *Pangaea*. Trapped near its centre, forced up on dry land, were remnants of oceanic crust and mantle (called *ophiolites*), and parts of the Eurasian/African Plate. These would form Newfoundland.

**Pangaea,
the Super-continent**
By about 300 million years ago the wandering continents had converged, forcing up a massive mountain chain that included the Appalachians.

Resistant Ridges

Beneath the great cliffs of the Long Range plateau lies the coastal lowland. There, tilted and shoved into place by continental collision, forested ridges of limestone rise a few metres above softer shale layers.

Newfoundland's Long Range Mountains were at the northeastern end of the Appalachians — heaved above the surrounding land by the force of continental collision.

On the coastal lowland of Gros Morne National Park, rock formations still show their heritage: they run southwest to northeast, at right angles to the path of the colliding continents.

© James Steeves

A New Ocean – The Atlantic

In TIME, THE PLATES RESUMED THEIR PONDEROUS TRAVELS.
Approximately 200 million years ago, Pangaea split 500
km east of the earlier Iapetus rift. No longer were the
rocks of Gros Morne at the edge of the North American
Plate. To the east was central Newfoundland — a portion
of crumpled Iapetus ocean floor material and remnants
of the volcanic island-arc. Farther east, a fragment of the
Eurasian/African Plate — today's Avalon Peninsula —
remained stuck tight to central Newfoundland as its
parent continent pulled away.

A new ocean, the Atlantic, grew between the
separating continents. It expands to this day as magma
wells up along the rift creating new ocean floor.
Bordering the rift is a massive structure called the Mid-
Atlantic Ridge. Separated by this great ridge, the plate
carrying North America moves away from the Eurasian
Plate at about three centimetres a year. For most of its
length this ridge is hidden deep beneath the sea. Its

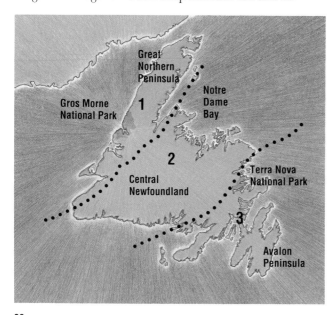

Three Zones of Newfoundland

1 - Continental margin of ancient
    North American Plate.
2 - Island-arc volcanic rocks and
    Iapetus Ocean sediments.
3 - Rock left behind when
    Eurasian/African Plate wrenched
    loose from Pangaea.

northern end (the Reykjanes Ridge) surfaces to form Iceland, with clusters of volcanos, hot springs, and geysers.

Gros Morne's rocks preserve an important part of eastern North America's early history. However, the park area has been above sea level for much of the time since the closing of the Iapetus Ocean. Instead of deposition occurring, for tens of millions of years the forces of erosion have eaten away at the rock. Water and ice wore the mountains down to a plain. As the weight of the great mountain chain was removed, the crust shifted slightly. Rising along faults, sections of the plain became the high plateau of the Long Range Mountains, and again erosion bit deeply into them.

The final shaping and polishing of the landscape took place during the last Ice Age, from about two million to ten thousand years ago. Ice flowed along previous river valleys and cut deep *fjords* such as Bonne Bay and Western Brook Pond. Boulders, gravel, and fine sand were dumped throughout the park and offshore when the ice melted.

On a smaller scale, geological processes shape the park today. Waves, currents, and ice sculpt coastal outcrops into sea stacks and caves, headlands, cliffs, and beaches. Cracked by ice and worn by running water, the Long Range Mountains and the Tablelands crumble year by year. Quiet gravity is always at work — slowly pulling down what plate tectonics has built up.

Exploring Gros Morne

Geological Map of Gros Morne

This simplified geological map of the park shows the positions of rock layers mentioned in the text.

ROCKS TRANSPORTED FROM THE IAPETUS OCEAN

Gabbro and other oceanic crust rock.

Peridotite — mantle rock.

Metamorphic sole — layer beneath the hot mantle rocks as they slid overland.

Volcanic rocks — remnants of seamounts and other volcanos.

ROCKS TRANSPORTED FROM THE ANCIENT CONTINENTAL MARGIN OF EASTERN NORTH AMERICA

Sandstone — deposited during rifting that produced the Iapetus Ocean basin.

Green sandstone — formed in deep water on top of the limestone breccia as the ocean closed.

Shale and limestone breccia — formed in deep water close to the foot of the continental shelf.

Mélange of folded scaly shales and blocks of other rock layers churned together as overlying rock slid westward on top of them.

IN SITU ROCKS OF THE ANCIENT CONTINENTAL MARGIN OF EASTERN NORTH AMERICA

Sandstone and shale — the top of the undisturbed rocks.

Limestone and dolomite — remnants of the thick carbonate bank.

Quartzite — the caps of Gros Morne, Big Hill and Killdevil.

Shale and limestone — early Iapetus Ocean floor sediments.

Rocks of the Long Range Mountains — about 1,250 million years old.

Cross Section of the Park

These cross sections show the structures that probably lie beneath the surface of the park along the lines A – B and C – D (at right). Coloured layers are equivalent to those on the map and on the diagrams on pages 20 – 25.

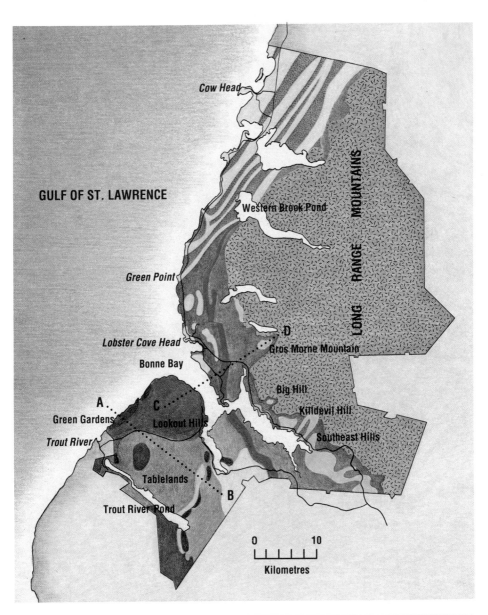

Cow Head

GULF OF ST. LAWRENCE

Western Brook Pond

Green Point

Lobster Cove Head

Bonne Bay

A · · · · · · · · C

Green Gardens

Trout River

Lookout Hills

Tablelands

Trout River Pond

B

D

Gros Morne Mountain

LONG RANGE MOUNTAINS

Big Hill

Killdevil Hill

Southeast Hills

0   10

Kilometres

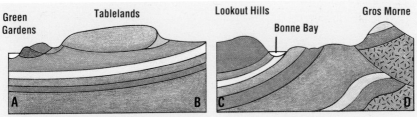

Green
Gardens

Tablelands

Lookout Hills

Bonne Bay

Gros Morne

A

B

C

D

Iɴ Gʀᴏs Mᴏʀɴᴇ Nᴀᴛɪᴏɴᴀʟ Pᴀʀᴋ, ʀᴏᴄᴋs ᴛᴇʟʟ ᴛʜᴇ sᴛᴏʀʏ ᴏғ a rift in the Earth's crust that became an ocean, of marine life that flourished on the continental shelf and in deep water, and of a collision of landmasses that obliterated the ocean and formed the super-continent Pangaea.

One of the strangest results of these events was the stranding of oceanic lithosphere on top of continental crust. The Tablelands in Gros Morne National Park and the complex of transported rocks below and around it have revealed much about the processes of plate tectonics and contributed to the park's designation as a World Heritage Site.

The Tablelands

Originally part of the mantle, many kilometres beneath the surface of the Earth, Tablelands rock lacks the nutrients to support plant growth and may be toxic to some plants.

© James Steeves

There are several ways to get more information about Gros Morne's rich geological history. At the Visitor Reception Centre at Rocky Harbour, rock specimens, exhibits, maps, reference material, and audio-visual programs explain the features and facilities of the park. Each summer, park interpreters offer guided walks and evening programs illustrated with slides and films. Viewpoint exhibits throughout the park interpret geology and other aspects of natural and human history.

Understanding the Rocks

Interpretive events such as this guided walk at Green Point help visitors to discover the geological significance of Gros Morne National Park.

Some of the park's rock formations are large and spectacular, while others are less obvious or are difficult to find. This section will help you to locate and understand some of the most interesting geological sites in Gros Morne National Park.

James Steeves

Cow Head

HOW TO GET THERE — Follow Highway 430 north from
the Visitor Reception Centre for 50 km, then take the
first road to Cow Head. Continue 3.6 km to the
intersection with Pond Road in the community of Cow
Head. Turn left and drive across a natural causeway onto

© James Steeves

Cow Head Peninsula; continue through the old village of Cow Head until you reach the government wharf. Park, and walk south along the shoreline. The breccia looks like grey cement with angular chunks of rock embedded in it.

WHAT YOU WILL SEE — All around the peninsula are layers of chunky grey limestone breccia. The angular fragments in the Cow Head breccia are pieces of shallow-water limestone. They tumbled from the edge of the carbonate bank that formed along the continental shelf 550 to 460 million years ago (see p.22). Some of the limestone chunks are several metres across. Layers of fine mud that blanketed the limestone have solidified into black shale.

The Cow Head breccias are important for their large size (coarseness) and thickness, and for the fossils that they contain. Here, geologists can compare shallow-water organisms (in the limestone) and deep-water organisms (in the shale). Cow Head rocks preserve a complete record of deposition from the Cambrian to the Ordovician. They provide one of the best opportunities in the world for learning about the different marine environments of that time.

Western Brook Pond

HOW TO GET THERE — Western Brook Pond trailhead is 27 km north of Rocky Harbour on Highway 430. From there, a 3-km trail leads to the western end of the pond. In season, a boat tour provides sightseeing cruises and access to the far end of the 16-km-long landlocked fjord. Check at the Visitor Reception Centre for departure times and fees. Allow at least half a day for this hike and cruise.

Landlocked Fjord

Cliffs at Western Brook Pond rise 650 m above the water, and extend another 165 m below.

Jeff Anderson

What you will see — The best way to explore this spectacular freshwater fjord is by boat. Glacially cut cliffs of granite, gneiss, and schist tower more than half a kilometre above the water. This Precambrian rock is some of the oldest on the Island of Newfoundland and forms the bulk of the Long Range Mountains.

Vertical seams as much as 10 m wide slash through the bedrock, reaching from deep in the Earth all the way to the cliff tops. The dark rock in the seams is *diabase*, which has the same chemical composition as basalt and gabbro. These *dykes* formed 600 million years ago as continents rifted and separated to produce the Iapetus Ocean (see p. 20). Continental crust thinned and cracked, then molten rock from below squeezed into fissures in the bedrock and cooled, forming sheet-like dykes. By dating the rock in these dykes, geologists have determined when continental rifting occurred.

Glacial excavation of an old river valley created the spectacular cliffs. They are constantly eroding, and are unstable. Large rocks occasionally fall from them and can pose a hazard. Do not approach cliff bases too closely.

Cliffs and Dykes

In the cliffs that line Western Brook Pond, erosion has exposed the edges of 600-million-year-old dykes. They formed when molten rock squeezed upwards into cracks in the crust and hardened.

Michael Wood

39

Green Point

HOW TO GET THERE — Follow Highway 430 north from
Rocky Harbour for 12 km. Turn left onto the gravel road
just past the Green Point Campground road. Park at the
pulloff in the field above the fish landing area, walk
downhill, and follow the beach north to the point. Please
respect the fishermen's property. Beware of falling rocks
— stand well clear of the cliff base.

© James Steeves

WHAT YOU WILL SEE — The rocks of the coastal terrace and cliffs at Green Point originated offshore on the continental rise of the Iapetus Ocean about 500 million years ago. These fossil-bearing ribbon limestones and shales were transported from about 100 km to the east of here during the same event that moved the Tablelands. Now they lie tipped at 115°, just past the vertical. This is one of the formations that geologists have studied in their attempt to reconstruct the ancient continental margin of eastern North America.

The prominent wall of limestone breccia here is the result of the same undersea avalanches that produced the rocks at Cow Head. This is a thinner layer with smaller fragments, and formed farther offshore. Green Point's greatest significance is that its layers of rock contain an intact sequence of fossils spanning the boundary between the Cambrian and Ordovician periods. This is one of the finest such sequences in the world. The boundary between the two periods lies somewhere in the notch just north of the breccia wall. Rocks of Cambrian age are to the south, Ordovician to the north. This site has been proposed as the world's best example (called a world *stratotype*) for this ancient time interval.

Layers of Time

Layered limestones and shales at Green Point were pushed into place relatively intact but were left almost overturned.

Gros Morne Mountain

HOW TO GET THERE – Gros Morne dominates the view from the Visitor Reception Centre and from the highway near Rocky Harbour. The James Callaghan Trail starts 3 km south of the Centre, on Highway 430. This 16-km trail leads to the top of the mountain and back. It is a rigorous seven-hour hike. A descriptive brochure is available from park outlets.

James Steeves

WHAT YOU WILL SEE – At 806 m, Gros Morne is the second highest peak in Newfoundland. This big isolated hill was a prominent landmark for early French fishermen.

A Resistant Cap

Highest point in the park, Gros Morne is the quartzite-capped hill after which the park is named.

Gros Morne Mountain is capped by shale, limestone, and quartzites, some of the earliest sediments laid down as the Iapetus Ocean widened (see p. 21). On a clear day, layers are visible on the northwestern face. The *quartzite* originated as quartz-rich sands eroded from Precambrian bedrock and deposited on an ancient tropical beach. Pressure from overlying layers of sediment fused the sand grains tightly together, producing the hard pink quartzite now exposed on the mountain. The iron mineral hematite gives this rock its colour.

Beneath its resistant cap, Gros Morne is formed of the same Precambrian gneiss as the Long Range Mountains. The valleys that isolate Gros Morne from the main range have eroded along faults that shifted long ago.

The Tablelands

HOW TO GET THERE – Take Highway 431 from Wiltondale to Woody Point. As you enter Woody Point, turn left and continue along Highway 431 towards Trout River. Follow the highway for 4.5 km, then turn left into the Tablelands Exhibit parking lot. This exhibit is about 11 km from Trout River.

James Steeves

Glacially Carved Valley

The U-shaped valley of Winterhouse Brook leads upwards to a high plateau, over frost-churned soil and ice-fractured rock.

WHAT YOU WILL SEE – The Tablelands is the most important geological feature in the park. This mountainous block is a slice of the rock that once lay beneath the Iapetus Ocean; it is a sample of the Earth's crust and upper mantle including the Mohorovičić Discontinuity, or *moho* — the transition zone between crust and mantle.

The yellow cliffs and eroded slopes of the Tablelands are peridotite from the upper mantle. These dense rocks are tens of kilometres above their usual position in the Earth's mantle, and their mineral composition is not conducive to plant growth.

Exposed surfaces of peridotite have weathered to a powdery tan colour, but freshly broken rock is dark greenish-brown. Water, under intense heat and pressure, altered some of the original minerals to *serpentine*, giving the rock a green snake-skin pattern. Serpentine is so characteristic of transported ocean crust that these rocks are called *ophiolites* after the Greek words "ophis" and "lithos" meaning "serpent stone". Oceanic lithosphere stranded on dry land is called an ophiolite suite.

Across the road from the barren Tablelands, and in sharp contrast to its yellow mantle rock, are the grey cliffs of the Lookout Hills — an enigmatic slice of ocean crust.

Green Gardens

HOW TO GET THERE – The parking lot and main trailhead are beside Highway 431, about 12 km from the turn-off to Trout River at Woody Point (3.5 km from Trout River). A four-hour walk takes you to the shore and back by the most direct route. Brochures for this trail are available at the Visitor Reception Centre.

WHAT YOU WILL SEE – The first part of the trail crosses the broad barren valley of Trout River Gulch, strewn with rocks carried from the Tablelands by glaciers. A fault runs the length of the valley beneath the rubble. South of the fault are the mantle rocks of the Tablelands; to the north are deformed and metamorphosed gabbro and other rocks from the ocean crust. At the coast, a panorama of lush vegetation and bold scenery betrays a change in the underlying rock.

Old Volcanos

Waves, currents, and ice have shaped the volcanic rock at Green Gardens into a coastline of sea stacks, cliffs, and caves.

Michael Burzynski

Along the shoreline at Green Gardens are sea stacks and caves eroded from the cliffs of basalt. The dark grey and reddish lavas flowed from Hawaiian-type volcanos in the Iapetus Ocean (see p. 22). These volcanic rocks were transported as one of the slices ploughed up by the closing ocean.

The basalt cliffs contain the bulbous outlines of *pillow lavas*, a feature typical of molten rock that cooled quickly underwater. They alternate with basalt flows fractured into *columnar joints*, usually the result of slower cooling in air. From this evidence, geologists have suggested that the lava was extruded very close to sea level.

Submarine Lava

Lava oozing from underwater fissures chilled rapidly and produced the rounded pillow lavas in this headland. Sandwiched by them is a band of air-cooled basalt flows.

The rocks here are rich in calcium, potassium, sodium, and aluminium, and a fertile soil has developed. These are veritable Green Gardens — natural meadows lush with grasses and other herbaceous plants — a traditional summer pasture for sheep.

Michael Burzynski

Trout River Pond

HOW TO GET THERE — From Woody Point, follow Highway 431 for 16 km to Trout River. Turn left through the community and continue 1 km to the park day-use area. The best view of the pond is from a lookoff along the road to the campground.

WHAT YOU WILL SEE — Trout River Pond, like Western Brook Pond, is a fjord-lake landlocked by the rebound of the land after the glaciers melted. The village of Trout River is built on delta gravels left by the melting ice. Marine shells in the terraces, and sea stacks stranded far above the waves attest to the relative fall of sea level.

Michael Burzynski

Looking up the pond, the barren, yellow peridotite cliffs of the Tablelands are a striking contrast to the low-lying forested land in the foreground. On the flank of the Tablelands are several hillocks of greyer rock, remnants of the oceanic crust that originally covered the entire top of the Tablelands. Over-steepened by glaciers, portions of the valley walls of Trout River Pond have slumped. Huge pieces of gabbro have slid down the sides of the Tablelands to their present positions. Plant growth helps to distinguish the gabbro from the peridotite.

Across the pond is the Elephant, a mass of grey gabbro that guards the Narrows. A boat trip to the upper pond reveals the sheer gabbro cliffs of North Arm Mountain on the south. A fault running the length of the pond separates North Arm Mountain from the Tablelands, but they are part of the same complex. North Arm Mountain displays the upper layers of ocean floor that have been eroded from the top of the upthrust Tablelands. At the inner end of the pond, a prominent ridge of metamorphic rock marks the sole of the ophiolite — the layer of rocks that was overridden by the hot mantle slab.

Earth's Displaced Mantle
Along Trout River Pond, both mantle and crust portions of an ancient ocean are exposed. The crustal material is grey and the mantle is yellow.

Lobster Cove Head

How to get there — Follow Highway 430 north from Rocky Harbour for 1 km. Turn left and follow the signs to the lighthouse. Several trails lead from the lighthouse down to the beach. Except at high tide, it is possible to walk completely around the headland.

© James Steeves

Mixed-up rocks

A beacon since 1898, Lobster Cove Head lighthouse rises from the twisted and crushed mixture of rocks called mélange.

WHAT YOU WILL SEE — This headland and Berry Hill are large chunks in the jumbled transported rock that forms the mélange. Beds of buff-coloured *dolomite* (magnesium-rich limestone), shale, and ribbon limestones were folded and faulted as slices of ocean floor slid over them (see p. 25). They were first deposited as the deep-sea portion of submarine landslides that produced the Cow Head breccia.

Interlayered with the limestones and shales are nodules and sheets of brown to green *chert*. This hard, glassy rock was formed by the precipitation of silica in deep water. Chert blades fashioned by inhabitants of this coast thousands of years ago are still sharp.

The coarse sandstone (*greywacke*) on which the lighthouse is built is younger, and contains sand-sized fragments of the Tablelands rocks, including black grains of chromite. These sediments were eroded from the above-water portions of the ophiolite as it approached from the east.

Southeast Hills

How to get there – The Southeast Hills contact is just uphill from the Southeast Brook Falls trailhead, beside Highway 430, about 24 km south of the park Visitor Reception Centre. From the parking lot, walk 200 m uphill along the highway. The outcrop is the roadcut on the left.

What you will see – Road construction has exposed a cross section through one of the most significant geological formations in the park. These tilted rocks display part of a contact zone, or unconformity, between rocks of Cambrian and Precambrian age. The 1,100-million-year-old gneiss of the Long Range Mountains (on the right) is directly overlain by a dark, pebbly sandstone deposited a mere 570 million years ago (on the left). This is not a spectacular site, but it is an important part of the park's geological story.

A 500-Million-Year Gap

A sample of the first sediments that settled to the floor of the Iapetus Ocean 570 million years ago, lies directly atop billion-year-old granite-gneiss.

Michael Burzynski

The Story Continues

Roadside lookoffs at the top of the
Southeast Hills provide sweeping
views of the surrounding hills and
valleys, which are even now slowly
returning to the sea in an endless
cycle of erosion.

James Steeves

Further Reading

THE GEOLOGY OF NEWFOUNDLAND AND THE APPALACHIAN region is currently an active field of research. Interpretation of Gros Morne's geology is subject to revision. An updated reading list of geological reports is available at the Visitor Reception Centre.

The Behavior of the Earth, Continental and Seafloor Mobility.
Allègre, Claude.
Harvard University Press,
Cambridge, Mass. 1988. 272 pp.

Exploring our Living Planet.
Ballard, Robert D.
National Geographic Society,
Washington, D.C. 1988. 366 pp.

The Dynamic Earth.
Scientific American special issue,
Vol. 249, #3, Sept. 1983.

Geological Report on the Area of Gros Morne National Park.
Williams, H.
Dept. of Earth Sciences,
Memorial University of Newfoundland,
St. John's Nfld. 1985. 40 pp.
Unpublished report available at the park.

Geological Maps

Geology, Humber Arm Allochthon, Newfoundland.
Williams, H. and P.A. Cawood.
Geological Survey of Canada, 1989,
Map 1678A, scale 1:250,000.

Geology, Topography, and Vegetation, Gros Morne National Park.
Berger, A.R., et al.
Geological Survey of Canada,
Miscellaneous Report 54, scale 1:150,000

These maps are available at the park bookstore or through
Geological Survey of Canada, Publication Section
601 Booth Street, Ottawa, Ontario, K1A 0E8

James Steeves

Index of Terms Explained